Dear Reader,

We're so glad you're reading our book. Like you, we love animals and are fascinated by the millions of creatures that share our planet. We hope you learn some amazing new facts that help you love them even more.

We also hope you'll feel inspired to take action to protect the animals you learn about. Maybe one day you will even work for one of the zoos or conservations we support!

However you choose to be there for the creatures we all care so much about, we want to be there for you every step of the way. Reach out to us and let us know what you love about animals, what you want to learn about them, and how you want to help them. We can't wait to hear from you!

Sincerely, *Jenny Curtis, Founder*

Dedicated to the incredible team at the Cheetah Conservation Fund, whose tireless efforts protect the fastest animal on earth.
—Jenny Curtis

For Wyatt, and his running adventures
-Allyson Randa

Author: Jenny Curtis

Designer: Allyson Randa

Photo Credits:

AdobeStock.com

Pixabay.com

Pexels.com

ISBN: 978-1-965081-10-5

This book meets Common Core and Next Generation Science Standards.

Table of Contents

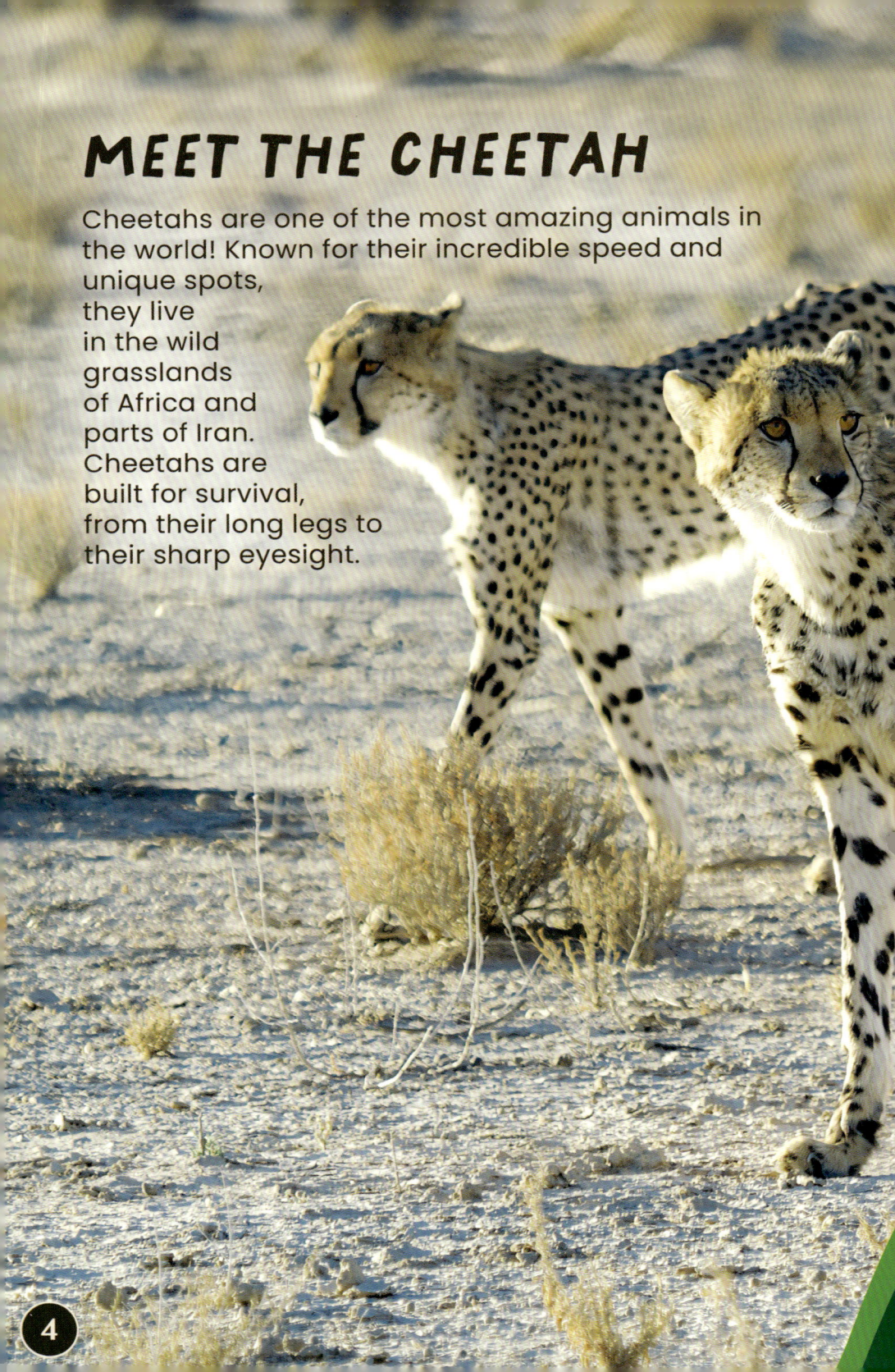

MEET THE CHEETAH

Cheetahs are one of the most amazing animals in the world! Known for their incredible speed and unique spots, they live in the wild grasslands of Africa and parts of Iran. Cheetahs are built for survival, from their long legs to their sharp eyesight.

Can you imagine what it would be like to sprint across a savanna as fast as a cheetah?

HOW FAST ARE CHEETAHS?

Cheetahs are the fastest land animals on Earth, reaching speeds of up to 70 miles per hour! Their long legs, flexible spine, and lightweight bodies make them nature's ultimate sprinters. Their claws act like cleats, and their long tails help them balance as they race after prey.

DID YOU KNOW?
Cheetahs can go from 0 to 60 miles per hour in just three seconds!

They spend more time walking than running to save energy.

Cheetahs stand out from other big cats like lions, leopards, and tigers. They're smaller and slimmer, designed perfectly for speed. Unlike lions, cheetahs are quiet hunters—they chirp and purr. They also don't climb trees like leopards. Instead, they use their speed and sharp eyesight to survive.

ADAPTATIONS THAT WOW!

Cheetahs have unique physical traits that help them survive. Their enlarged heart and lungs pump oxygen quickly, while their large nostrils and sinuses help them breathe efficiently. These adaptations are vital for powering their swift sprints and giving them an edge in the wild.

CHEETAHS CAN TAKE UP TO 150 BREATHS PER MINUTE DURING A SPRINT.

THE SECRET OF THE CHEETAH'S TAIL

Did you know a cheetah's tail acts like a steering wheel? When a cheetah runs at full speed, it uses its long, muscular tail to balance and make sharp turns. This is especially important during a chase when prey tries to zig-zag to escape. A cheetah's tail acts like a rudder, helping it stay balanced and turn quickly during high-speed chases. Speed and agility are only part of the story—cheetahs also rely on their sharp hunting skills to catch their next meal.

TAXONOMY

Kingdom: Animalia (An-i-MALE-yuh) All animals on Earth are in this kingdom.
Animalia: Living creatures.

Phylum: Chordata (core-DAH-tuh) This group includes all animals with a backbone, from fish to mammals.
Chordata: Having a spinal cord.

Class: Mammalia (muh-MAY-lee-uh) Mammals are warm-blooded, have hair or fur, and most give birth to live young.
Mammalia: Warm-blooded animals with fur.

Order: Carnivora (car-NIV-or-uh) Carnivores are meat-eating mammals, including cats, dogs, and bears.
Carnivora: Meat-eating mammals.

Family: Felidae (FEE-lih-day) This is the cat family, which includes lions, tigers, leopards, and house cats.
Felidae: The cat family.

Genus: Acinonyx (ASS-ih-NO-nix) Cheetahs are the only living species in this genus, which means "non-retractable claws."
Acinonyx: Cats built for speed.

Species: Acinonyx jubatus (ASS-ih-NO-nix joo-BAY-tus) This is the scientific name for the cheetah, meaning "maned or crested cat."
Acinonyx jubatus: The fastest land animal!

BUILT TO HUNT

Cheetahs are expert hunters, using their sharp eyesight and stealth to track prey during the day. They carefully stalk their targets, creeping closer and closer before making their move. After catching their prey, cheetahs eat quickly to avoid losing their meal to larger predators like lions or hyenas.

DID YOU KNOW?

Cheetahs are the only big cats that hunt mainly in the daytime!

In the savanna, nothing goes to waste. Hyenas and jackals often steal fresh kills, while vultures wait for their turn. Even beetles and insects help clean up the scraps!

WHY DO CHEETAHS HAVE SPOTS?

Cheetahs' spots aren't just for looks—they help them hide! Their coats blend into the tall grass, making it easier to sneak up on prey. Each cheetah has a unique pattern of spots, just like human fingerprints. Even their skin is spotted underneath their fur!

DID YOU KNOW?

Cheetahs prefer habitats with tall grass where they can hide from predators.

ROOM TO RUN

Cheetahs need large areas of open land to survive. Grasslands and savannas provide the perfect hunting grounds, with plenty of space for them to sprint after prey. However, as human populations grow, these habitats are shrinking. By protecting these spaces, we give cheetahs the chance to thrive and ensure other animals in the ecosystem are protected too.

TALKING WITHOUT WORDS

Cheetahs have many ways to communicate. They chirp to call their cubs, growl to show anger, and hiss to warn off danger. They also use scent marking, rubbing against trees or the ground to leave their 'signature.' These marks let other cheetahs know who's nearby!

DID YOU KNOW?

Cheetahs chirp when they're excited, which sounds like a bird.

WHY DO CHEETAHS PURR?

Unlike lions and tigers, cheetahs can't roar—but they can purr! A cheetah purrs when it's happy, relaxed, or bonding with its cubs. This soothing sound shows that cheetahs feel calm and content.

CHEETAH MOMS AND CUBS

Cheetah moms are amazing! They raise their cubs all on their own, teaching them how to hunt and stay safe.

Cubs are born with a fuzzy stripe of fur called a mantle that makes them look like honey badgers. This helps them stay hidden from predators.

MOTHERS MOVE THEIR CUBS EVERY FEW DAYS TO KEEP THEM SAFE.

Cubs stay with their mom for about 18 months before going off on their own. Cheetah moms work hard to teach their cubs important survival skills, but the cubs also learn a lot from each other.

TEAMWORK MAKES THE DREAM WORK

Once they're old enough to leave their mother, young siblings often stay together in small groups called "coalitions." Brothers, in particular, will stick together for life, hunting as a team to take down larger prey.

DID YOU KNOW?
Cheetah cubs learn through play!

Sisters take a different path. Once they leave their mother, they become independent, roaming alone. When it's time to raise cubs, they carefully choose safe dens and teach their young the survival skills they once learned.

FROM FLUFFY TO FIERCE

Cheetah cubs grow up fast! They're born blind and helpless, but within a few weeks, they begin exploring the world around them. At six months, they follow their mother and watch her hunt, learning important survival skills.

DID YOU KNOW?

Even though baby cheetah cubs are born small and slow, they start learning to run fast from a very young age! By the time they're about 3 months old, they begin practicing their incredible speed by chasing their mom's tail or playing with their siblings.

WHAT'S FOR DINNER?

Cheetahs are carnivores, eating mostly small to medium animals like gazelles, impalas, and hares. They also hunt warthogs, young wildebeest, and birds when given the chance.

DID YOU KNOW?

Cheetahs only need to drink water every three to four days!

DID YOU KNOW?

Cheetahs often eat only part of their prey, leaving the rest for scavengers.

NATURE'S BALANCE KEEPERS

Cheetahs are key players in the circle of life. By hunting herbivores like gazelles, they keep plant-eating populations in check. This helps prevent overgrazing, which can harm grasslands. Their leftover meals also feed scavengers like vultures and hyenas, ensuring nothing goes to waste. Cheetahs play an important role in keeping ecosystems healthy, but their daily lives are full of challenges.

A DAY IN THE LIFE OF A CHEETAH

Cheetahs spend most of their day resting and watching for prey. They are most active in the morning and late afternoon when it's cooler.

After catching a meal, cheetahs rest in the shade to recover.

DID YOU KNOW?

Cheetahs only hunt about once every 2-3 days to conserve energy.

Because they don't defend their territory like lions, cheetahs are always on the move to find safe places to live.

WHAT DO CHEETAHS DO AFTER DARK?

Unlike most big cats, cheetahs are diurnal, meaning they're active during the day. But what happens when the sun goes down? At night, cheetahs rest in grasses or under trees to stay hidden from predators like lions and hyenas. Their golden fur helps them blend into the shadows, keeping them safe until morning.

DID YOU KNOW?

At night a new set of animals comes to life. Lions, hyenas, and owls silently encompass the savanna.

CHEETAHS THROUGH HISTORY

Cheetahs have been part of human history for thousands of years. Ancient Egyptians kept cheetahs as pets and trained them for hunting. In India, rulers called them "hunting leopards." Today, conservationists and researchers work with cheetahs to learn more about them and find ways to protect them.

THE ASIATIC CHEETAH

Most cheetahs live in sub-Saharan Africa, mainly in savannas, open woodlands, and deserts. Did you know cheetahs also live in Iran? These are called Asiatic cheetahs, and they are a small, critically endangered population. They live in mountainous areas, where it's harder for them to find prey.

DID YOU KNOW?
Asiatic cheetahs have thicker fur to stay warm in the cool desert nights.

Conservationists are working hard to protect Asiatic cheetahs before they disappear forever. They face many unique challenges, but cheetahs everywhere are struggling to survive in the modern world.

CHEETAH HOMES

Cheetahs live in Africa and parts of Iran. They prefer wide, open spaces like savannas, where they can run fast without obstacles. Sadly, only about 6,500 cheetahs remain in the wild. Cheetahs' wide, open habitats are perfect for running—and for finding their favorite prey.

ARCTIC OCEAN

EUROPE

ASIA

IRAN

PACIFIC
OCEAN

AFRICA

INDIAN
OCEAN

AUSTRALIA

CTICA

WHY ARE CHEETAHS IN DANGER?

Cheetahs face many challenges, and their biggest threat is losing their habitat. As humans build farms and cities, cheetahs have fewer places to live and hunt. Farmers sometimes harm cheetahs to protect their livestock.

DID YOU KNOW?

Cubs are sometimes taken from the wild and sold as exotic pets, which further threatens wild populations.

A CHEETAHS FOOD CHAIN

Gazelles

Impala

Cheetah & Cub

Hares

Hyena

Lion

CHEETAHS HAVE BEEN AROUND
FOR ABOUT 3 MILLION YEARS!

PROTECTING LIVESTOCK

One of the best ways to protect cheetahs is by helping farmers protect their livestock. In Namibia, the Cheetah Conservation Fund trains special dogs called Anatolian shepherds and Kangal dogs to guard sheep and goats. These dogs scare off predators like cheetahs without hurting them.

DID YOU KNOW?

These dogs are friendly to humans but fiercely protective of their flocks.

This program helps farmers feel safer and reduces conflicts between humans and cheetahs. It's a win-win!

SAFE SPACES FOR CHEETAHS

Wildlife reserves and national parks are vital for cheetahs. These protected areas give them the space they need to hunt, rest, and raise their cubs. Thanks to protected areas, cheetahs have a fighting chance to thrive. By supporting these spaces, we're helping ensure their future.

DID YOU KNOW?
Protected reserves have helped cheetah populations stabilize in some areas.

Places like Namibia's Waterberg Plateau Park and Kenya's Masai Mara are some of the last strongholds for cheetahs in the wild.